Jernej Prišenk
Jernej Turk
Karmen Pažek

Optimization Processes in Agriculture

AF153151

Jernej Prišenk
Jernej Turk
Karmen Pažek

Optimization Processes in Agriculture

Optimization Processes on Agricultural Holdings with Multi-objective Programming Approaches

LAP LAMBERT Academic Publishing

Impressum / Imprint

Bibliografische Information der Deutschen Nationalbibliothek: Die Deutsche Nationalbibliothek verzeichnet diese Publikation in der Deutschen Nationalbibliografie; detaillierte bibliografische Daten sind im Internet über http://dnb.d-nb.de abrufbar.

Alle in diesem Buch genannten Marken und Produktnamen unterliegen warenzeichen-, marken- oder patentrechtlichem Schutz bzw. sind Warenzeichen oder eingetragene Warenzeichen der jeweiligen Inhaber. Die Wiedergabe von Marken, Produktnamen, Gebrauchsnamen, Handelsnamen, Warenbezeichnungen u.s.w. in diesem Werk berechtigt auch ohne besondere Kennzeichnung nicht zu der Annahme, dass solche Namen im Sinne der Warenzeichen- und Markenschutzgesetzgebung als frei zu betrachten wären und daher von jedermann benutzt werden dürften.

Bibliographic information published by the Deutsche Nationalbibliothek: The Deutsche Nationalbibliothek lists this publication in the Deutsche Nationalbibliografie; detailed bibliographic data are available in the Internet at http://dnb.d-nb.de.

Any brand names and product names mentioned in this book are subject to trademark, brand or patent protection and are trademarks or registered trademarks of their respective holders. The use of brand names, product names, common names, trade names, product descriptions etc. even without a particular marking in this works is in no way to be construed to mean that such names may be regarded as unrestricted in respect of trademark and brand protection legislation and could thus be used by anyone.

Coverbild / Cover image: www.ingimage.com

Verlag / Publisher:
LAP LAMBERT Academic Publishing
ist ein Imprint der / is a trademark of
OmniScriptum GmbH & Co. KG
Heinrich-Böcking-Str. 6-8, 66121 Saarbrücken, Deutschland / Germany
Email: info@lap-publishing.com

Herstellung: siehe letzte Seite /
Printed at: see last page
ISBN: 978-3-659-55318-9

UNIVERSITY OF MARIBOR

FACULTY OF AGRICULTURE AND LIFE SCIENCES

DEPARTMENT OF AGRICULTURAL ECONOMICS AND RURAL DEVELOPMENT

SCIENTIFIC MONOGRAPH

OPTIMIZATION PROCESSES IN AGRICULTURE

OPTIMIZATION PROCESSES ON AGRICULTURAL HOLDINGS
WITH MULTI-OBJECTIVE PROGRAMMING APPROACHES

JERNEJ PRIŠENK M.Sc.

FULL PROF. JERNEJ TURK Ph.D.

ASSOC. PROF. KARMEN PAŽEK Ph.D.

DEDICATION AND ACKNOWLEDGMENT

Agricultural economics science goes hand in hand with rural development processes and gives farmers necessary economic feedback. This connection was recognized by all members of the Department of Agricultural Economics and Rural Development in the Faculty of Agriculture and Life Sciences in Maribor in Slovenia. Until 2003, when the department was established, the members improved the interaction between these two scientific areas by solving unique rural development problems using new management decision programs. Management optimization processes, supported with simple usage programs on farms, can make positive contributions to the family farm's succession status forecasting in the future. Farmers can deal with different scenarios and investigate positive or negative results regarding future economic prospects. The optimization process is wide and complex, with several possible applications for the farms. Thus, the authors decided to represent one optimization approach implicated in two crucial management issues on the farms, with two different aims: feed ration optimization (horses' case study) to minimize costs and crop rotation optimization to maximize profits. The model's development process began in 2010, is now in the application process, and is based on department members' ideas. Acknowledgment also goes to other colleagues of the Department of Agricultural Economics and Rural Development, who supported this survey with professional comments and made positive contributions to the improvement of the represented model.

TABLE OF CONTENTS

LIST OF FIGURES

LIST OF TABLES

INTRODUCTION

Agriculture policy and its effects on agriculture and food production continue to change at a dangerous rate. In European Union countries, the food frequently comes from countries with lower labor costs, and consequently, homemade food products become exceptions on consumers' plates. The increase in the world's population leads to the production of a large quantity of food, and food production follows the trend to develop value-added food products with superior quality, convenience, availability, and affordability. Globalization food process issues require farmers to produce a maximum amount of heterogeneous food products with minimum input, but how they can do it? This isn't even the most complex problem for farmers. There is also the issue of food production in protection areas (i.e., less favoured areas— LFA areas or water protection areas). The issues related to food production are potentiated, especially in EU countries with a higher percentage of hills and mountainous regions covered with forests (i.e., Finland, Sweden and Slovenia), where technology in production systems is limited.

Individual farmers must repeatedly make decisions about what commodities to produce, by what method, in which seasonal time periods, and in what quantities. Decisions are made subject to the prevailing farm's physical and financial constraints and often in the face of considerable uncertainty about the planning period ahead. Already developed optimization techniques for planning the agricultural production are commonly unknown to farmers and/or they are presented as complex solutions which are often unacceptable in practice. To avoid this problem and help farmers solve these issues, it is crucial to present the optimization processes in understandable and easy ways. From this point of view, the aim of this book is to represent optimization methodology which can be used in the practice of farming and help farmers solve their optimization problems.

Deciding which optimization processes to use in agricultural can be described as multi-objective or multi-criteria decisions, because farmers need take many factors into account and solve several goals at one time. These kinds of problems often emerge when planning crop rotation and animal feed rations. Finding low-cost feeding options is a standard optimization problem (Brus et al., 2006) and can represent from 50% to 70% of total variable costs on a farm (depends on the production type). Žgajnar and Kavčič (2011) explained production planning as a complex task, which has to take into account the physical relations between input and output of natural farming resources along with cost-price rations and the farmer's preferences. Crop planning is related to many factors, both measurable and non-measurable (Mohamad and Said, 2011), and, according to Sharma et al. (2007), represents the most crucial factor of agricultural planning. Although crop planning and production depend on several resources, such as availability of land, water, labour, and capital (Sarker and Quaddus, 2002), they also impact a farmer's economic status.

The authors decided to introduce and represent two mathematical approaches implicated into two different fields. The methodology is derived from a (traditional) linear programming (LP) and weighted goal programming (WGP) approach. The combination of both methodologies represents the structure of the optimization models represented in this book. However, the LP approach has quite a long history. One of the earliest versions of an LP model was developed by Kantorovich (1960), translated from the Russian original, dated 1939, and more recently by Jolayemi and Olaomi (1995) and Hildreth and Reiter (1951), who applied LP in modelling crop rotation planning. The same mathematical technique was applied in the agricultural sector in Egypt by Bazaraa and Bouzaher (1981). Here readers can ask themselves why the authors favor the combination of two methods and consequently make the methodological approach more complex. The answer is quite simple—because of the shortcomings of LP, which have been defined by many authors (Tamiz et al., 1998; Rehman and Romero, 1984; Gass, 1987). The crucial defectiveness of LP is its

7

inability to optimize more than one goal (objective) at a time. To simultaneously solve more complex problems (i.e., combining crops, optimizing costs, and maximizing profits), LP does not represent the optimal choice, but there exists a (weighted) goal programming approach that is utilized for modeling and solving complex agricultural production planning problems (Mohamad and Said, 2011). The model was tested on real case studies for crop planning optimization and optimization of rations for sport horses.

The aim of this book is to represent the usage of linear and weighted goal programming to solve daily agricultural planning problems. The authors chose a step-by-step approach for developing an optimization model that presents the methodology as simply as possible. The book is structured in chapters that describe the methodology background, a description of the model's development process, and the implications and results of the model when it was tested in solving real-life agricultural problems.

CHAPTER 1: Linear and weighted goal programming: theoretical background

1.1 Linear programming (LP)

Linear programming (LP) is the process of taking various linear inequalities relating to a situation and finding the "best" value obtainable under those conditions. A typical example would be taking the limitations of materials and labor and then determining the production levels for maximal profits under those conditions. In real life, LP is part of a very important area of mathematics called "optimization techniques". This field of study (or at least the applied results of it) can be used daily in the organization planning processes of agricultural production systems. These real-life systems can have dozens or hundreds of variables or more. In algebra, though, you could only work with the simple (and graphable) two-variable linear case. The LP approach is commonly used in solving the following issues: standard maximum profit and minimum costs, diet for the animals, transportation planning and activity analysis.

However, this chapter is focused on representing case studies based on solving standard maximum and minimum issues in the agricultural production sector (i.e., crop rotation planning and optimization of feed rations). The objective for both issues is to try to find vectors for maximizing/minimizing the target/objective functions. In the case of crop rotation planning, the target function maximizes the profit on agricultural holdings, while in the case of feed rations, the farmers (also known as decision makers) focus on minimizing costs. The mathematical background of the maximum and minimum problems could be presented with equations from (1) to (5):

Maximum problem/Minimum problem

$$maxFR/\min C = c_1(fr_1)x_1 + \cdots + c_n(fr_n)x_n$$
(1)

Subject to the constraints:

$$a_{11}x_1 + a_{12}x_2 + \cdots + a_{1n}x_n \leq/\geq b_1$$
(2)

$$a_{21}x_1 + a_{22}x_2 + \cdots + a_{2n}x_n \leq/\geq b_2$$
(3)

...

$$a_1x_1 + a_2x_2 + \cdots + a_nx_n \leq/\geq b_n$$
(4)

and

$$x_1 \geq 0; \; x_2 \geq 0; \ldots; x_n \geq 0$$
(5)

Subject to:

$FR \text{ or } C$ = objective function (**F**armer's **R**eturn per crop rotation and **C**osts of feed rations)

fr_n = farmer's return of nth crop (€/ha)

c_n = costs of nth feed type at disposal (€/kg)

x_n = area of nth crop (ha)/the quantity of the nth feed type at disposal in the feed ration (kg)

a_{ij} = the quantity of the nth input of the nth crop/the quantity of nth nutrient in one unit of the nth feed type at disposal (i,j = 1…n)

b_n = available amount of the nth input/the amount of the nth daily nutritional requirements of different animal species

In some papers (Igwe and Onyenweaku, 2013; Visagie et al., 2004; Sinha and Sen, 2011; Žgajnar et al., 2009; Žgajnar et al., 2010), the mathematical formulation of LP is described in more detail and sometimes represents complex matrix approaches in different study fields. However, the structure could be represented in the easiest way by discussing each equation separately. The crucial point of the LP approach in agriculture optimization case studies is in equation (1), which represents the target function. It is defined as the sum product between costs and different feed quantitates or as the sum product between farmer's return and quantity of crops, including crop rotation in the case of crop planning. The main difference between minimum and maximum issues is in restrictions, where the sign between left- and right-hand side equations [(2) to (4)] is quite different. Sign " $\leq$ " is typical for solving maximum problems (maximization of farmer's return per crop rotation) and sign " $\geq$ " is typical for solving minimum problems (minimizing the costs of feed rations). Other restrictions represented by equation (5) are used to keep the results as non-negative values.

1.2 Weighted goal programming (WGP)

One common characteristic of the different kinds of management science models introduced so far (including LP, integer programming, and non-linear programming) is that they have a single objective function. This implies that all the managerial objectives for the problem being studied can be encompassed within a single overriding objective, such as maximizing total profit or minimizing total costs. However, weighted goal programming (WGP) provides a way to simultaneously strive toward several such objectives (input and output rations, farmer's preferences, and optimization of financial results on the farm), although sometimes it is impossible completely satisfy all goals. One of the main steps in the WGP mathematical technique is restrictions transformation from LP (see equations 2 to 4 with signs " $\leq$ or $\geq$ ") into goals (with sign " $=$ ") in WGP (equation 6). The basic approach is to establish a specific numeric goal for each of the objectives and then seek a solution that comes close to each of these goals. For solving multiple-objective problems, a system of weighting is commonly used to measure the relative importance of individual goals. Weights (w) are assigned to the objectives to measure the relative seriousness of missing their numeric goals. An objective function (equation 8) is formulated for each of the objectives. The overall objective is to minimize the weighted sum of deviations of these objective functions from their respective goals without minimizing or maximizing the goals themselves (Ferguson et al., 2006). One of the main issues within the WGP is incommensurability, especially where the units in one model are quite different (i.e., kg, g, l, ml). This could be solved with normalization techniques, such as those described by Tamiz et al. (1998). Within WGP, all deviations are expressed as a ratio difference (i.e., (desired − actual)/desired) = (deviation)/desired)) (Žgajnar and Kavčič, 2011). In this case, any marginal change within one observed goal is of equal importance, no matter how distant it is from the target value (Rehman and Romero, 1987). To keep deviations within desired limits, and to distinguish between different levels of deviations, a penalty function might be introduced into the WGP model (Rehman and Romero, 1984). This could be done when decision makers (i.e. farmers, local policy decision makers, and extension services) define over and under bounds for every interval. The mathematical background of the WGP, which was used for solving concrete agricultural planning problems, has already been represented in several papers (Žgajnar et al., 2010; Žgajnar and Kavčič, 2011; Zhang and Roush, 2002) and has the following mathematical formulation:

$$(a_1x_1 + a_2x_2 + \cdots + a_nx_n) + k_{1i}^- + k_{2i}^- - k_{1i}^+ - k_{2i}^+ = g_i$$

(6)

$g_i \neq 0$ and for all i=1 to n

$$(c_1(fr_1)x_1 + \cdots + c_n(fr_n)x_n) + k_{1i}^- + k_{2i}^- - k_{1i}^+ - k_{2i}^+ = C \; or \; FR \; from \; LP$$

(7)

C or FR $\neq 0$ and for all i=1 to n

$$minimize \; OBJ = s_1 \left(\sum_{i=1}^{l} w_i \times \left(\frac{k_{i1}^- + k_{i1}^+}{g_i} \right) \right) + s_2 \left(\sum_{i=1}^{l} w_i \times \left(\frac{k_{i2}^- + k_{i2}^+}{g_i} \right) \right)$$

(8)

$$k_{i1}^+ \leq pfp_{i1}^{max} g_i - g_i \tag{9}$$

for all i=1 to n

$$k_{i1}^- \leq g_i - pfp_{i1}^{min} g_i \tag{10}$$

for all i=1 to n

$$k_{i1}^+ + k_{i2}^+ \leq pfp_{i2}^{max} g_i - g_i \tag{11}$$

for all i=1 to n

$$k_{i1}^- + k_{i2}^- \leq g_i - pfp_{i2}^{min} g_i \tag{12}$$

for all i=1 to n

$$k_{i1}^+, k_{i1}^-, k_{i2}^+, k_{i2}^-, x_n \geq 0 \tag{13}$$

13

Subject to:

$k_{i1}^{+}, k_{i1}^{-}, k_{i2}^{+}, k_{i2}^{-}$ = positive and negative deviations of ith goal

g_i = goals (restriction from LP)

OBJ = objective function

w_i = weights of the ith goal

s_1, s_2 = penalty coefficients

$pfp_{i2}^{max}, pfp_{i1}^{max} > 1$ = penalty-function parameters defining the upper limit of the first and second intervals of the ith goal

$pfp_{i2}^{min}, pfp_{i1}^{min} < 1$ = penalty-function parameters defining the lowest limit of the first and second intervals of the ith goal

The objective function (*OBJ*), which minimizes the sum of all deviations, and all restrictions from LP, were transformed into goals, considering positive and negative deviations (equations 6 and 7). The objective function (equation 8) in WGP is defined as the sum product of weights and quotients between deviation variables and goals, which were multiplied with penalty coefficients. It is defined as the sum-product between the weights and the weighted deviations of the goals multiplied by the penalty coefficients (s_1 and s_2). The coefficients (s_i for i=1 to n) are important to keep the results during the intervals. The quotients between weights and penalty coefficient extend the function, as is shown in Figure 1. Equations (9), (10), (11), and (12) represent deviation restrictions, and equation (13) represents non-negative restrictions on deviations and decision variables.

14

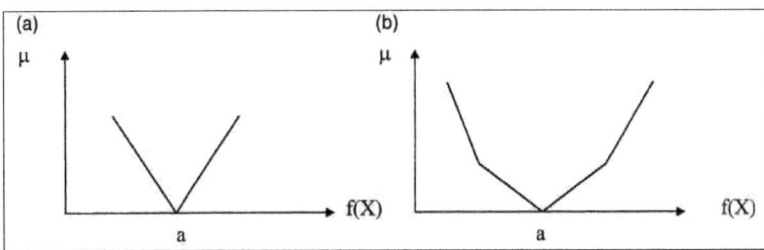

Figure 1: Scheme of a conventional V-shaped penalty function (a) and scheme of a four-sided penalty function (b) (Li and Yu, 2000).

CHAPTER 2: Model development process

An optimization tool (model) was developed in a Microsoft Excel framework that included the addition of "solver", which was used for the optimization process. Another addition that can be used is "WB—What's best Industrial", an add-in to Excel that allows the creation of large-scale optimization models. The authors decided to use only the solver to avoid additional problems connected with a lack of widespread use and other computer installation process issues related to WB, such as older versions of Microsoft Office not supporting the add-in. Solver is user-friendly tool and ready to use in all Excel framework, irrespective of the Microsoft Office version.

The model was created from two sub-models, where the first is based on the LP technique and the second is based on the WGP technique. The model was developed using the processes described in Žgajnar et al. (2010) and can be represented in four main steps (Table 1).

Table 1: Development steps of the optimization model

Step	Activity
1. Transforming data	Transforming restrictions and target functions from LP into goals in the WGP sub-model
2. Defining deviations and weights	Defining the deviations and the weights for separate goals from the first step
3. Defining restrictions	Defining the restrictions

	of the WGP sub-model
	(see chapter 1)
4. Defining target	Defining the target
functions	function (see chapter 1)

In the first step of model development, the previous restrictions from LP sub-models with signs " <= " or " >= " are transformed into the goals " = " in the WGP sub-model.

The goals can be satisfied partly or completely, depending on the deviation variables and the weights, which are defined during the second step of model development. The optimization model was tested using different scenarios, while the relative importance of the goals is quite different between the scenarios (controlling with different weight values). By defining the weights, decision makers create a priority list to satisfy the separate goals. The weight values can be between 0 and 100. The higher the weight, the higher the goals' relative importance on the goal "priority list" that is being set. During the second step, all deviation variables were also defined for each goal separately, and they represent the possible deviations in a positive and negative way from the goals. The deviations are represented as one tide (d_{1i}^- and d_{1i}^+) and two tide intervals (d_{2i}^- and d_{2i}^+) for each goal separately. With penalty intervals, farmers can create a flexible function and obtain several-sided penalty functions, such as those that were described by Rehman and Romero (1987) (see Figure 1).

After the second step, the crucial development process is a normalization process, which was used when different units were included in the model, described in the *weighted goal programming* sub-chapter. In a third step, restrictions for the WGP sub-model are defined with a non-negative sign " >= 0" and some other restrictions (penalty intervals) are defined for controlling deviations in the restricted size of the function. Finally, the last step is focusing on defining the target (objective) function, whose formulation is also represented in the *weighted goal programming* sub-chapter. The scheme of the optimization model is presented in Figure 2.

There are two different types of input data included in the optimization model. The first includes basic input data, such as farmer's return, mechanical labour, manual labour, fertilizer costs (optimizing crop rotation), nutritional value of available feed and food standards (optimizing feed rations). Basic input data are included in both sub-models, while the additional input data, such as penalty function intervals and weights, are suitable only for the WGP sub-model.

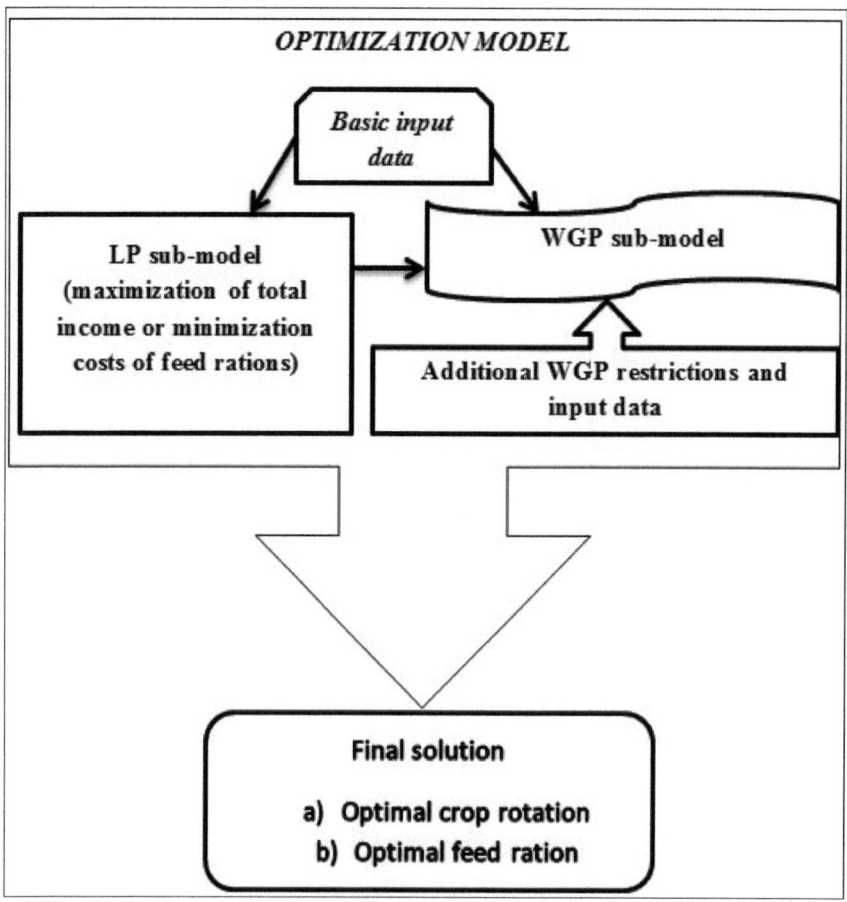

Figure 2: Scheme of the model for optimizing feed rations and crop rotation planning.

CHAPTER 3: Optimization of feed rations: a case study of the calculation of feed rations for sport horses

3.1 Optimization of feed rations

Unbalanced feed rations could be characterized as a twofold problem. First, underfeeding or overfeeding both cost money, but each situation can also have a negative impact on the environment. Overfeeding of some nutrients ultimately leads to an excess of unutilized nutrients, which can lead to pollution of soils and underground water. Both imbalances result in a determination of animal welfare, one of the concepts of cross-compliance that should be met by European Union (EU) farms to justify direct payment subsidies (Žgajnar et al., 2010). Poorly balanced feed rations certainly affect the welfare of animals and can worsen their health, especially those that are under constant psychological and physical pressure, such as sport horses, cows used for beef production, and daily dairy cows. Some studies (Zhang and Roush, 2002; Zimmermann, 2008; Žgajnar et al., 2010; Prišenk et al., 2013) have been conducted in the field of ruminant nutrition intended only for production purposes, so it was a research challenge to condense information on the development of mathematical methods to deal with animals that experience high levels of competitive stress. However, rations are most often constructed by experience, textbook-based knowledge, or by trial-and-error methods (by hand) (Žgajnar et al., 2009).

Calculating feed rations for sport horses is as complicated as calculating rations for production animals, because the quantity of feed provided can have cause positive and negative effects. Horse jumping is a sport that places physical and psychological burdens on a horse (Janžekovič et al., 2010; Prišenk, 2010; Prišenk et al., 2013). Consequently, it is very important to provide optimal feed rationing to satisfy all daily requirements. Assembling the appropriate feed rations for sport horses is

19

demanding and complex from a behavioral and nutritional perspective. Appropriate portions of the compounds are also very important from an economic perspective, because feed costs amount to up to 60% or more of total variable costs on a horse farm. Finding low-cost feeding compound combinations is a standard optimization issue (Brus et al., 2006). In this chapter, the calculation of winter and summer feed rations for sport horses is represented. The main difference is recognized in the daily nutrition needs which are dependent on training intensity. The main competition season is between May and the end of the September; during the winter, horses experience moderate or low training intensity.

3.2 Model input data

Model input data is data of the daily needs of horses with regard to nutrients and minerals, as well as the level of nutrients and minerals in each type of feed. This is represented in Table 2. The upper section of the table represents the summer and winter nutrition requirements of sport horses; the lower part represents the nutritive value of feed at disposal. Input data are relevant for sport horses with body weights between 550 and 600 kg, which represent the norm for active horses. The price of each feed at disposal represents retail prices in Slovenia and cost prices calculated with technological-economic simulations which represent the specific value of the unit price purchased. This value is used as a key factor in determining profitability, and in some stock market theories, it is used in establishing the value of stock holding.

Table 2: Food standards and nutritional value of available feed (according to Frape, 2010)

	FEED STANDARDS—WINTER				
	DM[1]	ME[1]	MP[1]	Ca[1]	P[1]
Weight of horse (550–600 kg)	14	126.2	1207	36	26
Unit	kg/day	MJ (megajoule)/day	g/day	g/day	g/day
	FEED STANDARDS—SUMMER				
Weight of horse (550–600 kg)	13.5	134.7	1448	44	32
Unit	kg/day	MJ/day	g/day	g/day	g/day
		NUTRIENTS				

Feed at disposal	DM	ME	MP	Ca	P	Price
Unit	g/kg	MJ/kg	g/kg	g/kg	g/kg	€/kg
Hay	860	7.4	80	2.9	1.7	0.16
Grass silage	350	5.6	62	6	2.2	0.07
Corn	880	14.2	85	0.2	3	0.18
Barley	880	12.8	95	0.6	3.3	0.16
Oats	880	12.1	96	0.7	3	0.17
Wheat	860	14.2	100	0.4	3.2	0.14
Endurix Cavalor	880	16.39	122	7.5	5.8	0.75

[1]Legend: DM = dry matter; ME = metabolisable energy; MP = metabolisable protein; Ca = calcium, P = phosphorus.

Additional input data shown in Table 3 includes restrictions, weights, and penalty coefficients, which are the most important part of the WGP sub-model. In the daily compilation of animal feed rations, it is necessary to consider the available quantity of each feed. This is why the second sub-model includes additional restrictions in the form of the relationship between calcium and phosphorus. The complex WGP also includes the ratio between calcium and phosphorus (Ca:P = 2:1), which is very important in horse nutrition. Otherwise, the proportion between Ca and P could have a strong negative effect on all occurrences in the organism. The feed ration must not exceed 5 kg of dry feed and 10 kg of grass silage and hay (Frape, 2010). The minimum value of dry feed is 2 kg per day and represents the minimum combination of grain maize, barley, oats, and wheat. Feed restrictions are relevant for both types of rations (winter and summer). As was discussed in previous chapter, the main difference between winter and summer rations is in signs. There is only one nutrient (DM) which has a mathematical sign " < " in LP, because it represents the maximum consumption capacity. Upper deviation is not possible because of the maximum consumption capacity, but if there are less DM in the diet, animals do not feel full.

The model was tested on four different scenarios (SC_1 and SC_2), two for winter and two for summer feed rations. These were dependent on different weight values to satisfy the different nutrient requirements shown in Table 3. In the first scenario, two main nutrients, metabolisable energy and metabolisable protein, have the highest values, while in the second scenario, the largest value of the weights belongs to costs of feed rations. To keep deviations within desired limits, and to distinguish between different levels of deviations, a penalty function might be introduced into the WGP sub-model (Rehman and Romero, 1984). Penalty function intervals are quite different and depend on the importance of the goal. For example, the most important nutrients in feed rations are ME and MP, and from this aspect, the intervals are very restricted as compared to other nutrients, such as DM, Ca, and P. Negative intervals for cost rations were not defined, as it is not necessary to limit goals if our main objective is to structure the rations so that we are optimizing costs and keeping them at a minimum. The penalty intervals for calcium and phosphorus are more extensive, because in this case, the proportion between them is more important, and this is defined under feed restrictions.

Table 3: Daily winter and summer needs of sport horses, shown as restrictions for the LP and WGP sub-models and additional input data (weights, penalty function intervals, and feed restrictions) for the WGP sub-model

	Unit	Winter		Summer		Penalty Function (PF)				Weights
						Interval 1		Interval 2		
		LP_W	WGP_W	LP_S	WGP_S	k_1^{-*}	k_1^{+*}	k_2^{-**}	k_2^{+***}	w_{sc1}/w_{sc2}
DM	(kg)	<14	14	<13.5	13.5	5%	0%	10%	0%	10/10
ME	(MJ)	>126.2	126.2	>134.7	134.7	3%	3%	8%	8%	100/50
MP	(g)	>1207	1207	>1448	1448	3%	3%	8%	8%	100/50
Ca	(g)	>37	37	>44	44	5%	5%	15%	15%	5/5
P	(g)	>26	26	>32	32	5%	5%	15%	15%	5/5
Cost	(€)	Costs from LP		Costs from LP		∞	10%	∞	20%	5/100

Additional feed restrictions

Type of restriction	Unit	Summer	Winter
Ca:P		2:1	2:1
Minimum dry feed	kg/day	2	2
Maximum dry feed	kg/day	5	5
Maximum hay + grass silage	kg/day	<15	<15

Legend: * k_1^-, k_1^+ = positive and negative deviations of the first interval; ** k_2^-, k_2^+ = positive and negative deviations of the second interval

3.3 Results and discussion

Table 4 shows the results for daily winter and summer rations for sport horses. The upper part of the table represents the structure of the daily rations, and the lower part represents the deviations of daily rations from daily requirements. It is critical to provide "light" rations for sport horses during the competition season. The rations should be balanced between roughage and feed that provides metabolisable energy and protein, which are especially needed during competition season (Salaić et al., 2010). Rations should also contribute to the animals' welfare without threatening their health (where 0.5 kg DM/day does not have a negative impact on the animal's health status) (according to Frape, 2010). The facts below (Table 4) show that for WGP, the quantity of hay and grain maize is less than what is provided with LP. This is important for two reasons: First is to provide "light" daily rations, and second is to keep rude albumen out of the rations, because this type of element can have a negative impact on the horses and cause their legs to swell. WGP_{sc1} includes the Endurix Cavalor into the rations in spite of its higher price, but the cost of the ration is the same as the LP ration. Generally, rations calculated with WGP are cheaper than LP rations, which confirms the assumption that when rations are not balanced, they have higher production costs per unit. This can be seen especially in the case of LP summer rations, where higher surpluses of megajouls, calcium, and phosphorus exists, which are also confirmed by deviations from the daily nutritional requirements shown in Table 4. Žgajnar et al. (2010) also noted that surpluses cost money not only in terms of increased pollution but usually also in higher private production costs. Finally, more balanced rations might also reduce GHG (Greenhouse gas) emissions

(Brink et al., 2001). WGP_{sc2} feed costs are lower (97.87€) for winter daily rations compare to LP (we assumed that the winter fattening period is between October and the end of April, which represents 210 days [approximately 30 days per month]). WGP_{sc1} feed costs are lower (68.04€) for the summer fattening period compare to LP. Also WGP_{sc2} feed costs are lower (78.42€) than for LP in the summer fattening period.

These results show that penalty intervals (also known as the penalty system) are crucial in controlling deviations of the metabolisable energy and metabolisable proteins which represent two important nutrient components in the rations. This statement has been confirmed with WGP_{sc1} results in both rations where deviations from daily requirements are quite small.

Table 4: Optimization results of winter and summer feed rations for sport horses

		Daily ration					
		Winter			Summer		
Type of feed	Unit	LP_w	WGP_{sc1}	WGP_{sc2}	LP_s	WGP_{sc1}	WGP_{sc2}
Hay	kg	7.59	10.00	10.00	8.76	12.52	12.41
Grass silage	kg	2.10	0.00	0.00	4.47	1.68	1.18
Corn	kg	1.30	0.00	1.23	1.30	0.97	0.71
Barley	kg	0.70	0.31	0.00	0.70	0.00	0.38
Oats	kg	1.75	2.84	1.08	1.75	0.85	0.96
Wheat	kg	1.25	0.00	0.77	1.25	0.61	0.69
Endurix Cavalor	kg	0.00	0.83	0.00	0.00	0.00	0.00
Cost	€/day	2.21	2.21	1.75	2.80	2.35	2.19
Cost deviation	%		0.00	-21.05		-16.17	-18.64
		Deviation from daily requirements					
DM	%	-16.73	-13.57	-19.18	0.00	0.00	0.00
ME	%	6.37	0.13	-8.45	15.97	0.12	0.12
MP	%	0.00	-0.11	-10.06	0.00	-8.06	-8.56
Ca	%	0.00	1.10	-18.08	24.13	7.74	0.93
P	%	26.82	20.59	1.57	25.57	1.27	0.57

3.4 Sensitivity test of the model

A sensitive test was used to determine the responsiveness, efficiency, and applicability of the developed tool. The test was based on changing values of the weights between costs and MP and ME (see Table 3; last column). Figure 3 shows the results of the summer ration costs; Figure 4 shows the winter ration costs. Because of the clear scheme of intervals, instead of the ∞, the bounds of k_1^- and k_1^- were defined as 100% of LP calculated costs.

In SC_1 of winter ration costs, where the weight is 5, the value of k_1^+ is outside of the bounds, while in SC_2, all values are inside the bounds. The higher weight values (100 in SC_2 winter ration costs) give better results from an economic standpoint as compared to SC_1, where the weight value is 5. The same results occurred when calculating summer ration costs. In the SC_2, the model calculated cheaper feed ration as compared to the SC_1. In general, the most favourable goal is for the weight to be higher than other goal weights. The model's sensitivity test explained the logical consequences of changing the weight values.

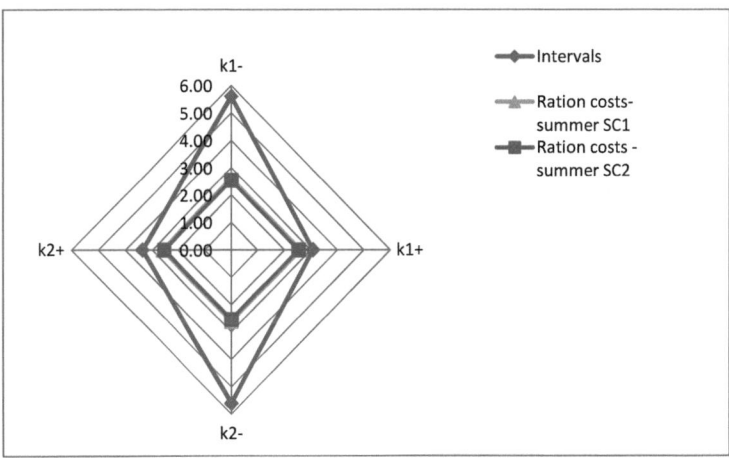

Figure 3: Deviations of summer ration costs from the penalty function intervals in both scenarios.

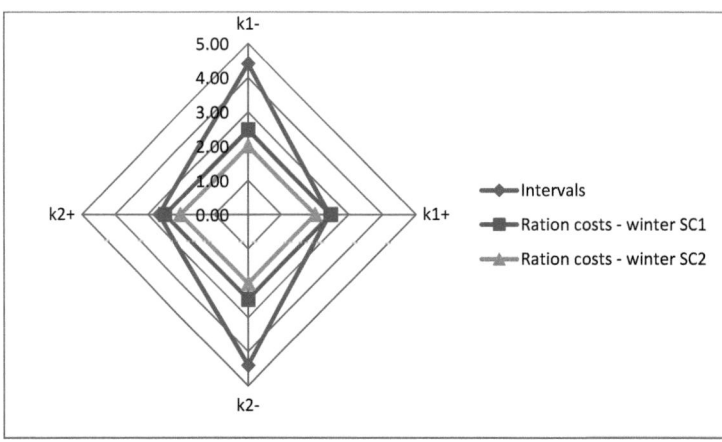

Figure 4: Deviations of winter ration costs from the penalty function intervals in both scenarios.

CHAPTER 4: Optimization of crop rotation: a Slovenian case study

4.1 Crop rotation planning

Crop rotation is a critical feature of all cropping systems, because it provides the principal mechanism for building healthy soils, a major way to control pests, and a variety of other benefits (Mohler and Johnson, 2009). Good crop rotation requires long-term strategic planning, not the monoculture approach, and it should increase Farmer's return, reduce debt and expenses, and expand the size of the farm. This multi-objective issue is often too complicated for farmers, and many of them are tempted to plant excessive acreage of their most profitable crop or to overuse certain fields for one type of crop (Mohler and Johnson, 2009).

In crop rotation, the planning process involves management and dealing with influences such as machinery, fertilizers, capital, labour, and other costs of production, which directly affect farm profitability. The decision issues, such as crop area planning, are based on conflicting and non-commensurable criteria to "satisfy" the desired outcome (Gupta et al., 2000), and expert farmers are continually balancing annual and multi-year (short- and long-term) decisions, which must be optimized for annual returns and cash flow. If the farmers are focused on maximizing profits or minimizing costs, the LP provides a sufficient technique; however, conventional mathematical programming schemes, such as LP, clearly cannot handle more objectives simultaneously (Mohaddes and Mohayidin, 2008), and the declaration of many goals at once is possible only with goal programming (Jafari et al., 2008).

In this chapter, the authors present the case study of a Slovenian arable farm. The average size of Slovenian farms is 6.5 to 6.7 hectares, which is quite small compared to farms in other EU countries. This represents the greatest challenge for the developed model; it must be accurate so that decision makers can use it to optimize crop planning on large farming systems abroad. The model covers three main areas: size of the farm, costs, and profit.

4.2 Input data

The input data were taken from the *Calculations Catalogue for Management Planning on Farms in Slovenia (CMPS)*, which provides the most relevant and publicly accessible data. Seven different crops were included in the model, with calculated farmer's return, mechanical labour, manual labour, and fertilizers costs per hectare (Table 5). These data are relevant for both sub-models, although Table 6 shows input data, such as weights, penalty function intervals, and restrictions that are relevant only for the WGP sub-model. The authors assumed that the available cropping area for the model is 6.5 hectares, which is the average size of a farm in Slovenia and also represents the restriction area for cropping. Cost restrictions were calculated as a sum of all potential available crops in the crop rotation, but they did not represent the universal form of calculation and could be changed per farmers' requests. Other resources of the input data include a technological-economic simulation of a total-costs calculation for different crops on the farms, but this could be different from farm to farm, especially in terms of fertilizers costs, because of the different soil structure and soil exhaustion on the farms. To avoid these problems, input data and restrictions were taken from the CMPS (2011). The technological-economic simulations will be good data resource when the model is used on specific agricultural holdings.

Table 5: Input data for both sub-models (Source: CMPS, 2011)

Crops	Farmer's return (€/ha)	Mechanical labour	Manual labour	Fertilizers	Cropping area (ha)
		COSTS (€/ha)			
Corn	1632.0	158.5	126.5	295.5	? LP
Grass silage	1294.0	128.2	225.5	422.7	? LP
Buckwheat	865.0	106.6	60.5	83.3	? LP
Barley	982.0	138.4	115.5	206.6	? LP
Wheat	1082.0	138.8	115.5	230.9	? LP
Potato	5036.0	470.7	891.0	404.3	? LP
Canola	1132.0	131.3	82.5	250.0	? LP
Restrictions		1273	1617	1893	6.5

The weights have different values for the three scenarios (see Table 6). In the first scenario (SC$_1$) the authors tried not to exceed the requirement cropping area (6.5 ha) while also satisfying the other goals (weight for cropping area is 100). In SC$_2$, the farmer's return – profit was maximized with the weight 100. The term "maximize" of the farmer's return was used instead of the term "to keep in bounds" because there is no underachievement of this goal (the sign for k_1^+ and k_2^+ is ∞). In the last scenario, the authors tried to keep all costs in bounds, independent of the farmer's return and cropping area. Tiny intervals were defined for the cropping area (3% on the first level and 8% on the second level), because of the limited area of arable land on the farm. Other intervals are more extended, and this is important from the aspect of function limitation (see sub-chapter *1.2 Weighted goal programming*) and consequently for controlling deviations in the restricted size of the function. From the restriction aspect, the crucial difference between LP and WGP is in signs (" < ") in LP and (" = ") in WGP.

Table 6: Additional input data for the LP and WGP sub-models

Goals		Restrictions for both sub-models		Penalty function/PF				
				Interval 1		Interval 2		Weights
	Unit	LP	WGP	k_1^-	k_1^+	k_2^-	k_2^+	$w_{sc1}/w_{sc2}/w_{sc3}$
Farmer's return	€/ha	n.a.	= result from LP	5.00 %	∞	10.00 %	∞	**10/100/5**
Mechanical labor costs	€/ha	< 1272.50	= 1272.50	5.00 %	5.00 %	15.00 %	15.00 %	**5/5/100**
Manual labor costs	€/ha	< 1617.00	= 1617.00	5.00 %	5.00 %	15.00 %	15.00 %	**5/5/100**
Fertilizer costs	€/ha	< 1893.30	= 1893.30	5.00 %	5.00 %	15.00 %	15.00 %	**5/5/100**
Cropping area	ha	< 6.50	= 6.50	3.00 %	3.00 %	8.00 %	8.00 %	**100/10/5**

4.3 Results and discussion

The results of the optimization tool can be seen in Table 7, which is structured in three parts. The first part shows the economic results of the crop rotation in the LP and WGP models. The goal amounts are represented separately. The second part of the table shows the structure of the crop rotation in hectares, and third part shows the deviations from the restrictions (expressed in percentage values). Deviations were calculated based on the restrictions represented in Table 5.

Table 7: Results of the model represented as crop rotation structure, costs, and deviation from restrictions

Goals / restrictions				WGP	
	Unit	LP	SC$_1$	SC$_2$	SC$_3$
Farmer's return	€	12140.43	12254.62	12596.12	12140.43
Mechanical – labor costs	€	1272.50	1272.50	1272.50	1269.01
Manual – labor costs	€	1617.00	1617.00	1617.00	1617.00
Fertilizers costs	€	1614.50	1893.30	1893.30	1893.30
Cropping area	ha	6.35	6.50	6.89	6.50
Crop rotation structure					
Type of crops	**Unit**				
Corn	ha	1.00	1.69	1.80	2.37
Grass silage	ha	0.00	0.56	1.02	0.67
Buckwheat	ha	1.00	0.07	1.51	0.00
Barley	ha	0.11	0.85	0.00	2.47
Wheat	ha	2.00	1.14	0.00	0.00
Potato	ha	1.24	1.07	1.06	0.99
Canola	ha	1.00	1.12	1.50	0.00
Deviation from restriction					
Goals	**Unit**	**LP**	**SC$_1$**	**SC$_2$**	**SC$_3$**
Farmer's	%	n.a.	1.00	4.00	0.00

return					
Mechanical – labour costs	%	0.00	0.00	0.00	0.00
Manual – labour costs	%	0.00	0.00	0.00	0.00
Fertilizers costs	%	-15.00	0.00	0.00	0.00
Cropping area	%	-2.00	0.00	6.00	0.00

Results of the model show the differences between the analyzed sub-models in formulated crop rotation. In WGP SC_1 and SC_2, the farmer's returns were higher as compared to the LP sub-model results, as well as the cropping area, which contains the weight issues. In SC_1, the higher weight values were defined for the cropping area, and in SC_2, for the farmer's return. From the results, we can see a connection between the cropping area in use and the farmer's return. The weight 100 for the cropping area in SC_1 contributes to the satisfaction of the cropping area restriction (6.5 ha) and consequently results in an income increase of 1% (114.19 €). The difference could be seen in SC_2, where the higher weight value was defined for the farmer's return, and the consequence is 4% extra profit (455.96 €) for farmers, as compared to the LP result. In this case, 0.39 ha represents the extra used cropping area, resulting in an additional 6% of hectares. However, this is in the defined second bound k_2^+, which was defined on the 8% (see Table 6). From the SC_3 results, we see another connection between costs and farmer's return. In SC_3, the costs had the highest value and consequently the farmer's return cannot deviate from the defined goal (12140.43 €). The results from the LP sub-model are completely unacceptable, because they do not satisfy the recommended cropping area in use (6.5 ha), which automatically restricts the farmer's return and profit for the farmers. The most profitable, and for the farmers the most acceptable, results are calculated in the WGP_{SC2} scenario, because the farmer's return outweighed the fertilizers costs (for 278.80 €), which were calculated by the LP sub-model.

4.4 Sensitivity test of the model

The sensitivity test of the model is represented in the same way as for the calculation of feed rations. It is based on the changing weight values among farmer's return, cropping area in use, and costs. In the second scenario of WGP, the weight for farmer's return maximizes the results, and all other goals are inside the bounds. The responsiveness and efficiency of the model confirm the results for SC_1 and SC_2, where the proportion of the weights of farmer's return and cropping area is changed from 10:100 in SC_1 to 100:10 in SC_2. These proportions are shown in Figures 5 and 6. Figure 5 shows the deviations of farmer's return, while Figure 6 shows the deviations of the cropping area. Instead of the ∞ (also used in the feed ration calculation), the farmer's return bounds k_1^+ and k_2^+ were defined as 100% of LP farmer's return, because of the clear scheme intervals represented in Figure 5. This sensitivity test also explained the logical consequences of the changing weight values, because in both cases, the highest weight values have priority over the other lower values. If we review both figures together, we can see that a combination of the proportion 100:10 between farmer's return and cropping area provides a better result and also keeps all values inside the defined bounds. In SC_1, the weight of the cropping area outweighed the weight of farmer's return. In this case, the model firstly satisfied the cropping area goal and then maximized the farmer's return as much as possible on the 6.5 ha. It did not allow us to exceed the 6.5 ha cropping area, which was defined as a restriction of the model (see Table 5).

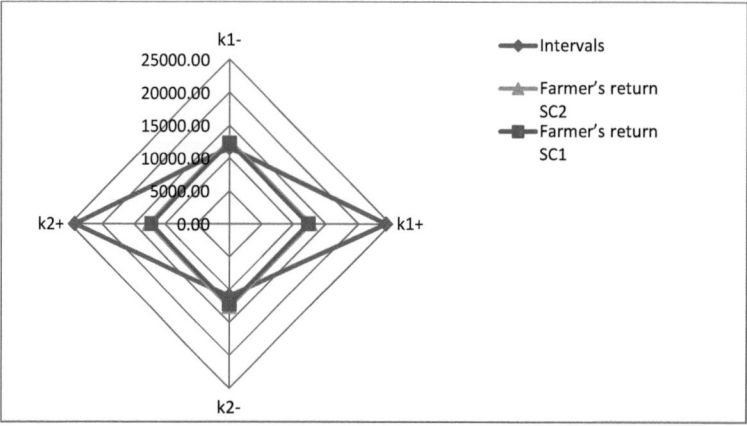

Figure 5: Deviations of farmer's return from the penalty-function intervals in scenarios 1 and 2.

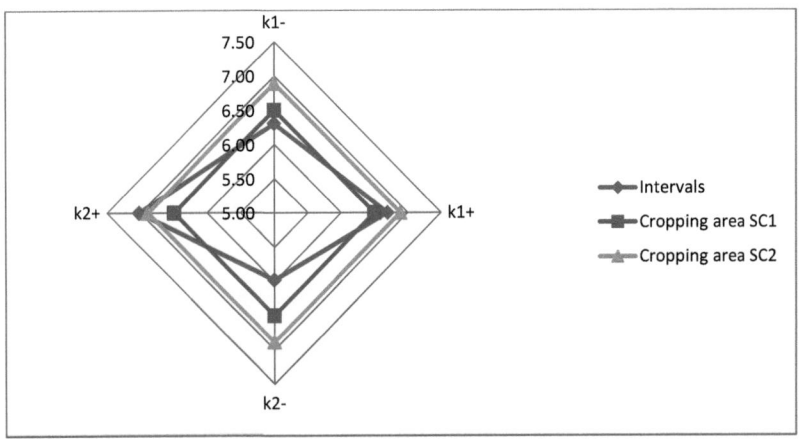

Figure 6: Deviations of cropping area from the penalty-function intervals in scenarios 1 and 2.

CHAPTER 5: LP and WGP sub-model framework: graphical representation of the models

This chapter focuses on the geographic design of the LP and WGP sub-models, as shown in MS Office Excel for easy viewing. It also discusses how to create the optimization tool and understand the mathematical background of the model represented in Chapter 1. Both sub-models have the same (universal) structure, independent of the subject of solving (crop rotation, feed ration optimization, planning fertilization process, or other optimization issues on agricultural holdings). Figure 7 shows the LP sub-model framework, and Figure 8 shows a portion of the WGP sub-model framework for optimization crop planning.

The crop rotation structure in the LP sub-model is represented by cells I4 to I10, as shown in Figure 7. The target function, which defines the farmer's return, is represented by cell M2. The mathematical formulation of the LP target function can be seen in the window fx. The restrictions are represented by cells E12, F12, G12, and I12, nonnegative restrictions by cells between E13 and E19, and additional restrictions with $<=$ signs between cells J14 and J17. Results of the goals in the LP sub-model are represented by H14 to H17.

A more complex structure can be seen in Figure 8, which represents a portion of the WGP sub-model for optimization crop planning. In this case, the crop rotation structure is shows in cells I5 to I11 and results of the goals in cells L28 to L32. The target function in the WGP (see cells K6 and L6) is designed with a mathematical formulation in the fx window, supported by penalty coefficients (K10 and L10) and an intermediate mathematical formulation process, which does not appear in this part of the WGP scheme. Penalty function intervals with percentage description are represented by cells F17 to F21, G17 to G21, H17 to H21 and I17 to I21. Cells E13, F13, G13, and I13 represent the restrictions from Table 5 and 6, and additional restrictions are represented by cells B26 to H26 and continue below.

The feed ration structure in the LP sub-model is represented by cells J20 to J26, as shown in Figure 9. The target function, which defines the ration costs, is represented by cell N38. The mathematical formulation of the LP target function can be seen in the window *fx*. The restrictions are represented by cells K15, L15, M15, N15 and O15, nonnegative restrictions by cells between J30 and J36, and additional restrictions with <= and >= signs between cells O30 and O34. Results of the goals in the LP sub-model are represented by Q20 to Q26.

A more complex structure can be seen in Figure 10, which represents a portion of the WGP sub-model for optimization feed rations. In this case, the feed ration structure is shows in cells P18 to P24 and results of the goals in cells J29 to J34. The target function in the WGP (see cell Q36) is designed with a mathematical formulation in the *fx* window, supported by penalty coefficients and an intermediate mathematical formulation process, which does not appear in this part of the WGP scheme. Penalty function intervals with percentage description are represented by cells M29 to M34, N29 to N34, O29 to O34 and P29 to P34. Cells J40 to R40 represent the additional restrictions which continue below.

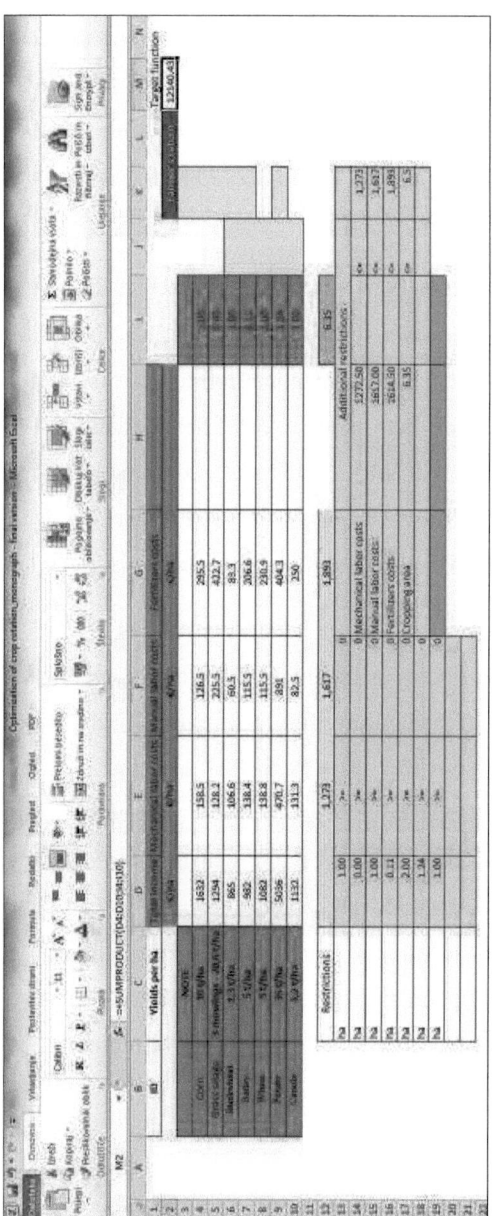

Figure 7: LP sub-model framework in Microsoft Office Excel (2010) for crop rotation planning.

40

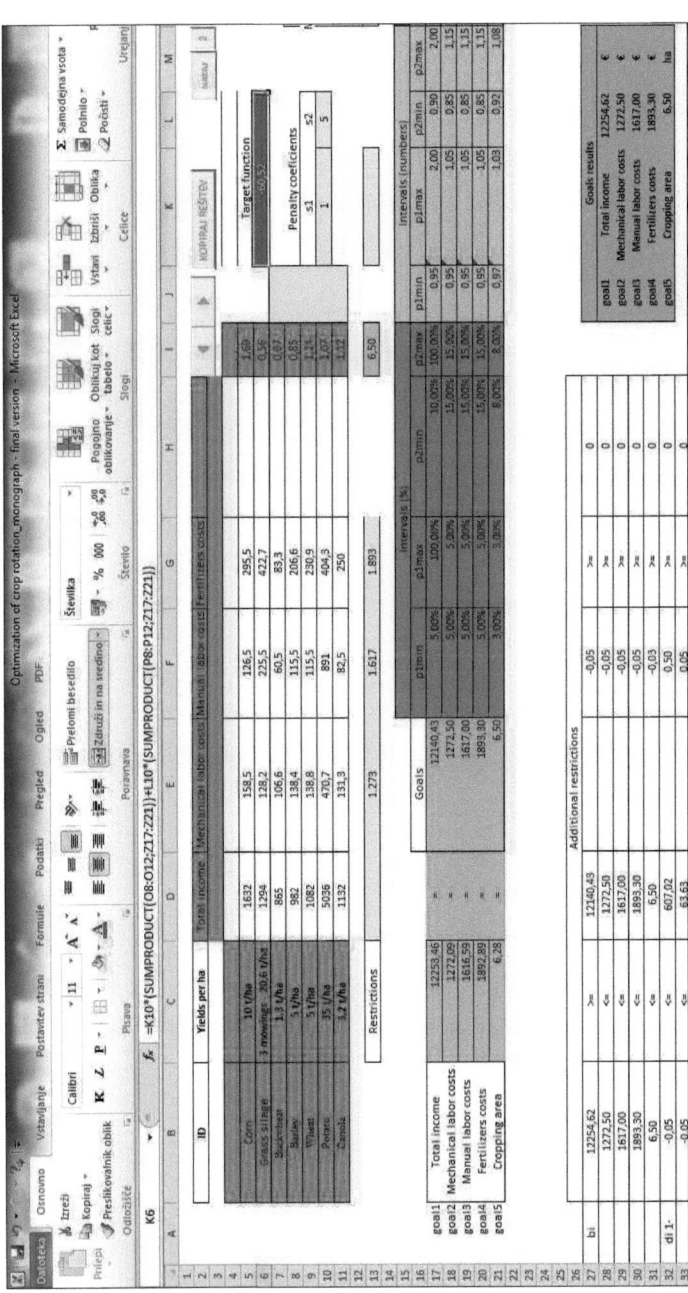

Figure 8: A portion of the WGP sub-model framework in Microsoft Office Excel (2010) for crop rotation planning.

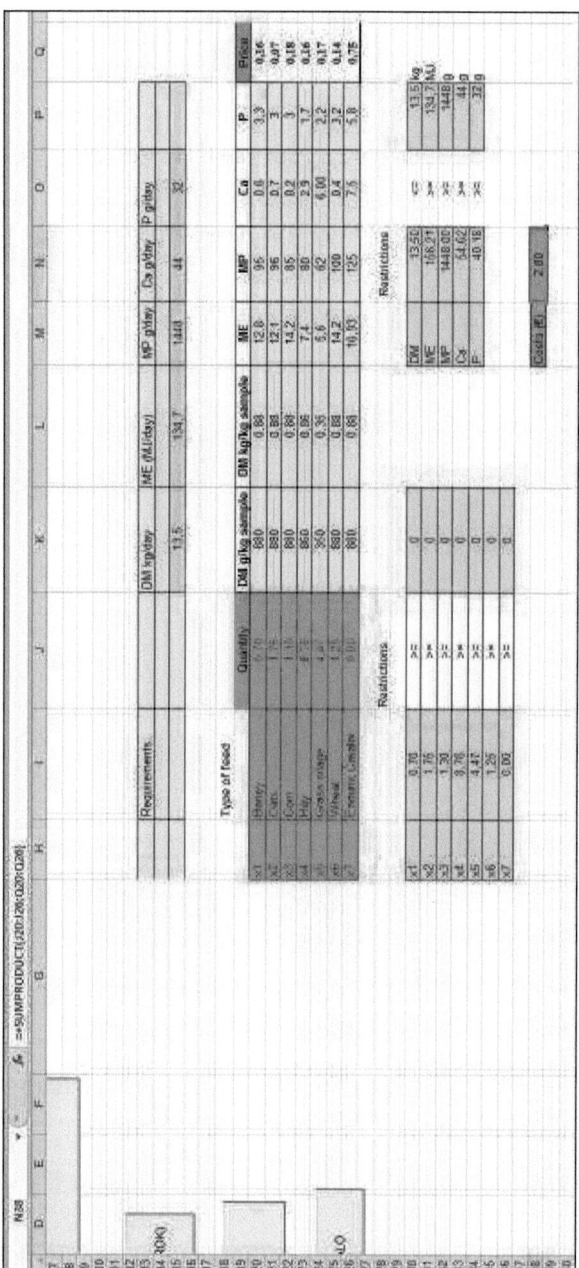

Figure 9: A portion of the LP sub-model framework in Microsoft Office Excel 2010 for optimizing feed rations (a case of the summer feed ration).

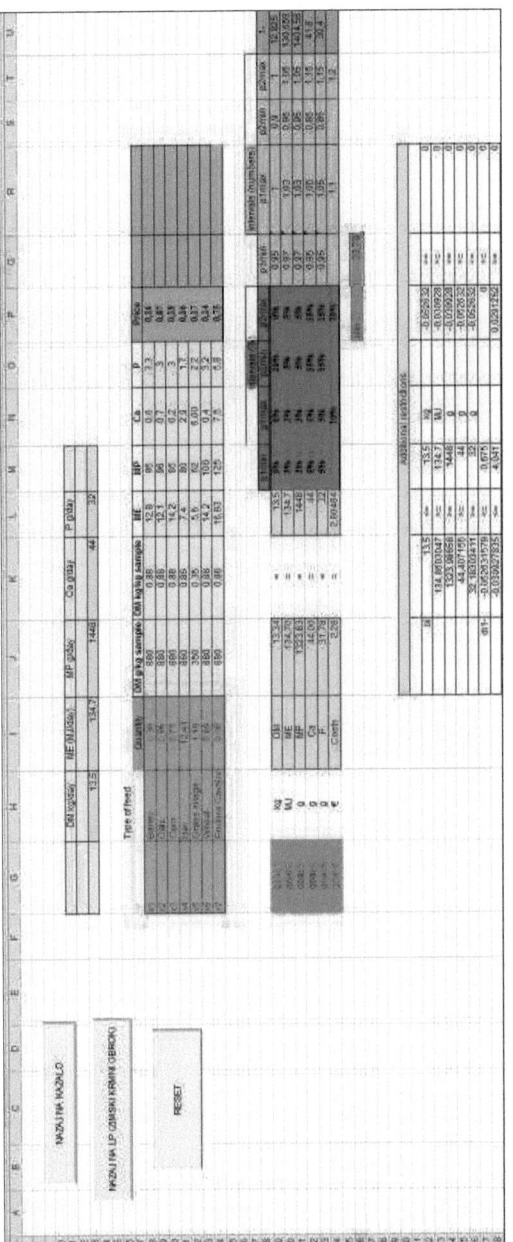

Figure 10: A portion of the WGP sub-model framework in Microsoft Office Excel 2010 for optimizing feed rations (a case of the summer feed ration—WGP$_{sc2}$).

43

CONCLUSION

The complexity of the mathematical formulations in solving decision-making problems often alienates potential users, especially when they must purchase special tools. The represented methodological technology in this paper can be used by all farmers, because it rests on a widely available tool—Microsoft Office Excel.

With knowledge of the cropping area, daily nutrition needs for different animal species, and other calculated results, farmers will be able to demonstrate different scenarios in the context of risk management. The tool's results can address some staple issues in pursuing crop rotation:

- What cropping strategies should be used and how much land should be allocated?

- What amount of extra resources (mechanical and manual labour, fertilizer, etc.) should be used?

- What proportion of the crops should be harvested and sold?

- What amount of the fertilizer could be saved to create crop rotation in the following year?

The model expresses a wide range of practical applicability and could address a host of other issues and goals (e.g. financial results and also economic and environmental parameters, such as soil erosion). The model was tested on a relatively small (cropping) area in Slovenia, and results from sensitivity test confirm its responsibility. Thus, it is assumed that the model will provide relevant results on large farms. The main advantage of the model was confirmed with a sensitivity test and represents an opportunity for farmers to improve their results inside the bounds of their defined goals. Upgrading the model with specific fertilizer requirements can help farmers in fertilization and management planning on the farm. Calculating the crop rotation in the second year and years after could be also upgraded, depending on farmers' requirements, such as including new organic crops or new restrictions. Therefore, the results indicate that making changes in crop rotation, as suggested by the WGP sub-model, facilitates an increase in farmers' profits.

The tool's results for optimizing feed rations can address some advantages and staple issues in pursuing the calculation of those feed rations:

-   The model improved the individual element of feed.
-   Surpluses of Ca and P calculated with the LP sub-model may mean a significant deterioration in animal health and an increase in the total costs of feed ration.
-   Minimizing surpluses also reduce the costs of veterinary services.
-   Optimization processes can assist livestock farmers in determining what amount and types of crops they need and how this translates to feed ration. This information can be included as additional restrictions in the model for crop rotation planning.

In practice, it is impossible to prepare the rations by including precisely 0.31 kg of barley, 0.61 of wheat, and so forth, but these amounts can rounded to one decimal. Using this kind of "robust" data will not have a crucial influence on the higher price of rations. The added value of the represented methods can be recognized in the way that the decision maker is allowed to introduce any number of additional requirements that will be considered when calculating new output results.

Because of its ease of use, the model is suitable for a wide range of users, but in particular, it could be used by local policy decision makers and extension services. The model is an open program where farmers can create (with weights and penalty functions) a priority list of their goals, indicating the most important tasks on their farms.

SUMMARY

A multi-objective programming model was developed to solve crop rotation planning issues with the objectives of maximizing farmer's return and minimizing production costs. The model was also used for optimization (minimization) of feed costs for horses. It is structured from two different mathematical approaches. The first sub-model is based on the linear programming (LP) method, while the second is based on the weighted goal programming (WGP) method with penalty functions. The results indicate that the WGP provides better solutions from an economic standpoint. WGP also provides better solutions for diversified and economically feasible crops that are included in a crop rotation.

BIBLIOGRAPHY

Jernej Prišenk, M.Sc. achieved his M.Sc. at University of Maribor, Faculty of Agriculture in 2012. He is active as assistant for agricultural economics and rural development at the Department of Agriculture Economics and Rural Development on the Faculty of Agriculture and Life Sciences, University of Maribor, Slovenia. His research includes development of multi-criteria models (using multi criteria decision analysis) for the assessment of local food supply chains, development of nonparametric models using linear programming and weighted goal programming for optimization feed rations for different animal species and optimization production processes on farm holdings. His research field is also econometric modelling implicated in different agricultural surveys. He is involved in researching development of mountain rural regions in Slovenia (different types of food supply chains) and works as a researcher in multiple national and international research projects. Jernej Prišenk is the author and co-author of a bulk of scientific papers with SCI and JCR factors. His full bibliography is available at: http://izumbib.izum.si/bibliografije/Y20140527160826-34865.html.

Full Prof. Jernej Turk, Ph.D. holds a position of Full Professor in Agricultural Economics and Farm Policy. He is currently the Head of Department of Agricultural Economics and Rural Development at the Faculty of Agriculture and Life Sciences of the University of Maribor till its establishment. He was a Dean of the faculty between 2003 and 2011. His research interest is devoted to agricultural economics modeling for specific policy purposes. He is the author and co-author of a bulk of scientific papers with SCI and JCR factors. His full bibliography is available at: http://izumbib.izum.si/bibliografije/Y20140527161115-10036.html.

Assoc. prof. Karmen Pažek, Ph.D. achived her Ph.D. at University of Maribor, Faculty of Agriculture in 2006. She is active as Associated Professor for Farm management on the Department for Agricultural Economics and Rural Development on Faculty of Agriculture and Life Sciences, University of Maribor. Her research includes development of decision support tools and systems for farm management (simulation modeling, multicriteria decision analysis, option models) and economics of agricultural production. She is involved in teaching activities as the supervisor at postgraduate study programs and involved in national and international research projects. She is author or coauthor numerous of scientific papers with impact factor. Her full bibliography is available at: http://izumbib.izum.si/bibliografije/Y20140527160952-22514.html.

REFERENCES

1. Bazaraa, M.S., Bouzaher, A., 1981. A Linear Goal Programming Model for Developing Economies with an Illustartion from the Agriculture Sector in Egypt. Manag. Sci. 27(4), 396-413.

2. Brus, M., Rozman, Č., Janžekovič, M., 2006. Program application. What`s best industrial in mathematical optimization of beef ra¬tion (in Croatian). In: S. Kapun, T. Čeh, I. Ambrožič (Editors).

3. Brink, C., Kroeze, C., Klimont, Z., 2001. Ammonia abatement and its impact on emissions of nitrous oxide and methane in Europe – Part 1: method. Atmospheric Envir. 35, 6299-6312.

4. CMPS, 2011. Calculations catalogue for management planning on farms in Slovenia, Jerič, D. (Eds.), Caf, A., Demšar-Benedičič, A., Leskovar, S., Oblak, O., Soršak, A., Sotlar, M., Trpin, Š.D., Velikonja, V., Vrtin, D., and Zajc, M., Chamber of Agriculture and Forestry of Slovenia, Ljubljana, pp. 267.

5. Ferguson, E.L., Darmon, N., Fhmida, U., Fitriyanti, S., Harper, T.B., Premachandra, I.M., 2006. Design of optimal food-based complementary feeding recommendations and identification of key "Problem Nutrients" using goal programming. J. Nutr. 136 (9), p. 2399-2404.

6. Frape, D.L. (Editor), 2010. Equine Nutrition and Feeding. 4th Edition. Wiley-Blackwell, Chichester, pp. 496.

7. Gass, S., 1987. The setting of weights in linear goal-programming problems. Comp. Oper. Res. 14, 227–229.

8. Gupta, A.P., Harboe, R., Tabucanon, M.T., 2000. Fuzzy multiple-criteria decision making for crop area planning in Narmada river basin. Agr. Syst. 63, 1–18.

9. Hildreth, E., Reiter, S., 1951. On the choice of a crop rotation. In: Koopmans, T ed. Activity analysis of production and allocation. John Wiley and Sons. Pp. 177–188.

10. Igwe, K.C., Onyenweaku, C.E., 2013. A Linear Programming Approach to Food Crops and Livestock Enterprises Planning in Aba Agricultural Zone of Abia State, Nigeria. Am. J. Exper. Agric. 3(2), 412-431.

11. Jafari, H., Koshteli, Q.R., Khabiri, B., 2008. An Optimal model using goal programming for rice farm. Applied Mathematical Sciences. 2(23), 1131–1136.

12. Kantorovich, L., 1960. Mathematical methods of organizing and planning production (translated from a report in Russian, dated 1939). Manag. Sci. 6, 366–422.

13. Janžekovič, M., Prišenk, J., Muršec, B., Vindiš, P., Stajnko, D., Čuš, F., 2010. The art equipment for measuring the horse`s heart rate. J. Achiev. Mater. Manuf. Eng. 41, 180–186.

14. Jolayemi, J.K., Olaomi, J.O., 1995. A mathematical programming procedure for selecting crops for mixed-cropping schemes. Ecolological Modelling. 79(1), 1–9.

15. Li, H.L., Yu, C.S., 2000. Solving multiple objective quasi-convex goal programming problems by linear programming. Intl. Trans. In Op. Res. 7, 265-284.

16. Mohaddes, S.A., Mohayidin, M.G., 2008. Application of the fuzzy approach for agricultural production planning in a watershed, a case study of the Atrak watershed, Iran. American-Eurasian J. Agric. Environ. Sci. 3(4), 636–648.

17. Mohamad, N.H.J. and Said, F. 2011. A mathematical programming approach to crop mix problem. Afr. J. Agric. Res. 6(1), 191–197.

18. Mohler, C.L., Johnson, S.E., 2009. Crop rotation on Organic Farms: A Planning Manual. Ithaca, New York: Natural Resource, Agriculture, and Engineering Service.

19. Prišenk, J., 2010. Heart rate monitoring by non-invasive method in sport horses and development of the linear model for calculation of the horse fodder. B.Sc. Thesis. University of Maribor, Faculty of Agriculture and Life Sciences, pp. 64.

20. Prišenk, J., Pažek, K., Rozman, Č., Turk, J., Janžekovič, M., Borec, A., 2013. Application of weighted goal programming in the optimization of rations for sport horses. J. Ani. Feed Sci. 22, 335-341.

21. Rehman, T., Romero, C., 1984. Multiple-criteria decision making techniques and their role in livestock ration formulation. Agr. Syst. 15, 23–49.

22. Rehman, T., Romero, C., 1987. Goal programming with penalty functions and livestock ration formulation. Agr. Syst. 23, 117–132.

23. Zhang, F., Roush, W.B., 2002. Multiple-objective (goal) programming model for feed formulation: An example for reducing nutrient variation. Poultry Sci. 81, 182-192.

24. Zimmermann, A., 2008. Optimization of sustainable dairy-cow feeding systems with an economic-ecological LP farm model using various optimization processes. J. Sustain. Agr. 32 (1), 77–94

25. Salaić, J., Baban, M., Domaćinović, M., Mijić, P., Sakač, M., Bobič, T., Budimir, K., 2010. Comparison of the nutritional value of feed rations for sport horses in harness discipline (in Croatian). Stočarstvo 64, 29-39.

26. Sarker, R.A., Quaddus, M.A., 2002. Modelling a nationwide crop planning problem using a multiple criteria decision making tool. Comp. Ind. Eng. 42, 541–553.

27.   Sharma, D.K., Jana, R.K., Gaur, A., 2007. Fuzzy goal programming for agricultural land allocation problems. Yugo. J. Operat. Res. 17(1), 31-42.

28.   Sinha, B., Sen, N., 2011. Goal programming Approach to Tea Industry of Barak Valley of Assam. Applied Mathematical Sci. 5(29), 1409-1419.

29.   Tamiz, M., Jones, D., Romero, C., 1998. Goal programming for decision making: An overview of the current state-of-the-art. Eur. J. Oper. Res. 111, 569–581.

30.   Visagie, S.E., De Kock, H.C., Ghebretsadik, A.H., 2004. Optimising an integrated crop-livestock farm using risk programming. Orion. 20(1), 29-54.

31.   Žgajnar, J., Erjavec, E., Kavčič, S., 2010. Multi-step beef ration optimi¬sation: application of linear and weighted goal programming with a penalty function. Agr. Food Sci. 19, 193–206

32.   Žgajnar, J., Juvančič, L., Kavčič, S., 2009. Combination of linear and weighted goal programming with penalty function in optimisa¬tion of a daily dairy cow ration. Agric. Econ. 55, 492–500

33.   Žgajnar, J., Kavčič, S. 2011. Weighted goal programming and penalty functions: whole-farm planning approach under risk. EAAE 2011 Congress, Zurich, Switzerland 30th August–2nd September 2011. Pp 1–11.